BI BOOKS

A Bite-Sized Public Affairs Book

You're on Mute!

Optimal Online Video Conferencing - in Business, Education & Media

Edited by

Dr Alex Connock

Cover by

Dean Stockton

Published by Bite-Sized Books Ltd 2020
©Alex Connock 2020

Bite-Sized Books Ltd Cleeve Road, Goring RG8 9BJ UK
information@bite-sizedbooks.com
**Registered in the UK. Company Registration
No: 9395379**

ISBN: 9798658820378

For Caroline and Michael

Contents

Introduction

You're On Mute!

Optimal Online Video Conferencing - in Business, Education & Media

Dr Alex Connock

Summary

This book is a Zoom lens into the explosive, viral spread of online video conferencing - which impacted almost every segment of society and the global economy during Coronavirus in 2020. From Facetimes for critically-ill intensive care patients, to Hollywood TV shows streamed from iPhones, what best-practice has been discovered, and what is coming in the near future ?

Billions of people on video

Colombian novelist Gabriel Garcia Marquez said: "Everyone has three lives: a public life, a private life and a secret life." The dramatic shift to video conferencing, necessitated by the global pandemic from March 2020, brought all of those three lives instantly into play.

Communication changed for billions of participants, in millions of business, schools and government departments, in hundreds of countries all around the world. The consequences - sociological, technical, financial, and personal - are still developing, and largely still uncalculated.

Zoom came to be the category-defining brand – but was actually late to the video conference party, by almost a decade. Zoom's power derived from what The Economist described as the platform's "intuitive simplicity of an Apple product." An interface that did not require the 'hour's induction session' required by some other tools, enabled it to reach critical scale and exploit the network effect, by being pre-loaded at each end of myriad global communications. Launched in 2012, it topped 300 million global users per day by April 2020.

But Zoom was far from alone. Group video chat on WhatsApp in Italy also increased by 1000% in a week in April 2020. Skype still had scale in 2020. Launched back in 2003, bought by Microsoft for $8.5 billion in 2011, it had been given an unsuccessful software makeover and was possibly on a journey to merge with the newer Teams, in the same stable. Add in Google Classroom, the high-spec Blue Jeans and countless other Edtech and video conferencing platforms – many of which are discussed in this book - and the growth amounted to a decade's evolution in the technology and practice of communication, in a matter of weeks.

Private and Secret lives

At the most basic level, in video conferencing the private became public. This global, crowdsourced body of work alleviated some of the stress of coronavirus, the servers becoming a priceless historical record of the background and foreground life of the world's working population.

"My boss turned herself into a potato on our Microsoft Teams meeting, and can't figure out how to turn the setting off," tweeted Rachele Clegg in Washington DC of her boss Lizet Ocampo, the political director at People for the American Way. "So she was just stuck like this the entire meeting." Being stuck in potato mode: perhaps the quintessentially harmless video conferencing war story – amongst a global, anecdotal, digital patchwork quilt of many thousands.

Paolo Longo, parish priest of the Church of San Pietro and San Benedetto di Polla in Italy, activated face filters by mistake while preaching Mass at the altar via a Facebook Live session – to find his head decorated in gangster hat and glasses, confetti and pink dumbbells at the altar. This was just one incident amongst many, which – true to Garcia Marquez' vision of the public, private and secret - also included clandestine lovers revealed behind business calls, unfortunate background paintings, suspect books about Nazis on shelves behind politicians, undressed public officials conducting civic works meetings (in multiple countries) and silent movie-style background pratfalls. These were trivial, chimeric disasters – a busy social stream of tales in a global village brought together by near-instant immersion in video conferencing. Of one such incident, the UK's ITV political editor Robert Peston wrote: "On a bleak day, this will improve things."

Public lives

Then there were the turns of the sociological wheel towards a new and glittering digital future. Never had people focussed on their likeness as much as in the pandemic. In the UK, a Manchester orthodontics consultant reported a dramatic surge in enquiries from adults requiring major cosmetic dental work - after seeing their likeness on Zoom for hours on end each day. Hundreds of writers met every morning at 11 to share their likenesses in an hour-long silent, global videoconference writing session:: "It helps my concentration," said one.

In cuisine, London restaurant Deli Cat & Sons ran virtual supper clubs, with chef Karan Ghosh sending meal and cocktail ingredients to participants' homes (tandoori salmon, crumble with cardamon), and collectively prepping in real time. Across the Atlantic, at Harvard, Nina Gheihman defended her doctoral dissertation via Zoom on "Veganized: How Cultural Entrepreneurs Mainstreamed a Movement", just as Zoom itself mainstreamed as a global cultural movement.

In Hollywood, perennial home of status-oriented thought leadership, a new organisational nuance was introduced into the democracy of the flat mosaic of faces on Zoom or Teams. Thoughtful creatives from shows like *Arrow* and *Parks and Rec*, ran a Zoom Room for artsy website Ars Technica to benchmark the dynamic creative changes taking place around TV production, describing a hierarchy of meetings where scene-specific discussions were held in breakouts, with only the all-powerful showrunner sitting in each. But the novelty wore off fast: "There is less joking around. We did every possible Zoom joke we could do in that first week," said Executive Producer Dan Goor of *Brooklyn Nine-Nine*. Zoom led to more focussed work days, fewer bathroom breaks, less "losing the room to endless joke offs," and fewer complex team-building routines like *Wheel of Fortune*-style takeout selection processes.

Each facet of the media viewed Zoom through its own prism: Business Insider offered tips of recognising "Zoom burnout" and handling it; GQ offered an experiential account of "spiritual dowsing" via Zoom to handle "negative emotion and sexual healing" via a Soho-based therapist, Sezen. The writer said that after the call: "I did not feel anything. Or more correctly, I can't tell if I felt anything."

The era of out-of-office did not mean the end of office politics - where through starvation of the social cues of in-person interaction, executives became overtaken by isolation, and compensated with presenteeism. Insecure feelings were exacerbated by misreading of emails. The Financial Times (June 18, 2020) quoted a relatable marketing executive: "You're riddled with paranoia. If I don't hear back from someone I'm worried." Andre Spicer, professor of organisational behaviour at the UK's Cass Business School said virtual working can give space for that fear to grow. "We don't have the usual informal feedback to make sense of what is going on." Richard Kramer at Stanford described "hypervigilance," the scrutinising of peers and bosses for meaning. And Carnegie Mellon's Anita Woolley said: "If someone doesn't respond to an email, you think: 'What does that *mean*?'

Zooming into trouble

Video conferencing didn't just change office politics. It changed real politics too, catching governments, activists, security specialists and spies in its widening net.

The British government held press conferences on Zoom – with requisite technical difficulties, participants accidentally left on mute, and awkward follow-ups occasionally silenced, one hopes by accident not design. The government signed a £2m contract with Zoom on 7 April, part of 731 Zoom licences purchased, including by the Ministry of Defence –about two weeks before the National Cyber Security Centre suggested politicians and civil servants should maybe stay off the platform for sensitive discussions.

In fact, security considerations threatened Zoom's primacy. On April 9, the US Senate told members not to use Zoom because of concerns about data security. Zoom apologised to China for not censoring international calls with participants from that country, and shut down meetings. Reportedly it disabled the US-based accounts of a former Tiananmen Square leader, Wang Dan and his team mates, and was quoted as taking this position: "When a meeting is held across different countries, the participants within those countries are required to comply with their respective local laws".

The Financial Times speculated that that 'relevant law' may have been the 'catch-all' Chinese test of "provoking quarrels."

Later in June 2020, Zoom adjusted course and converted to end-to-end encryption for all customers, not just those who pay for a subscription, set to start in July. As the Guardian reported, Zoom CEO Eric Yuan had previously said they planned to exclude free calls from end-to-end encryption to make sure it is still possible to "work together with FBI, with local law enforcement in case some people use Zoom for a bad purpose" - an implicit apparent acknowledgement that it could at least in theory monitor content of calls. Even on Zoom's July relaunch, reported Tech Crunch, end-to-end encryption would not be activated by default — but offered as an option. This was because - Zoom said - encryption limits some meeting functionality "such as the ability to include traditional PSTN phone lines". It remains to be seen whether Zoom's privacy shift for all accounts, including free ones, would be enough to maintain its stunning lead.

Some big numbers

Wikipedia is not a reliable research tool – but its transparent edits can be a short-cut, anecdotal measure of the pace of change. Type in 'Comparison of web conferencing software', and you will see a fascinating – and remarkably fast-evolving - real-time comparison of dozens of work-from-home tools and their shared and unique functionalities. The pace of edits is accelerating. In fact, the taxonomy of the remote work market overall is exponentially growing. A map[1] from Signal Fire divides it into 14 clear market segments like Hiring, Knowledge Management and Communication, and then dozens of sub-categories, like Virtual Whiteboards (Miro, Mural), Project Management (Trello, Asana) and of course Video Conferencing (Zoom, Google Meet, Daily.co, Around, Jitsi and Whereby, to select but a few.) In all, just on this one indicative chart, there are over 200 substantial companies in the remote working space.

Money is being made. Across early Covid, Zoom's revenue soared. As the Economist noted, one analyst report started with "Wow", another with "Holy Cow," praising CEO Eric Yuan's ability to pivot rapidly and learn from his mistakes (like around security). Defying the instant and deliberately government-mandated global lockdown-driven recession, Zoom's revenue jumped 169% to $328m in the three months to end of April 2020, with free cash flow of $250m. In percentage terms, only Amazon could match that lockdown bonanza. The news temporarily drove the stock market value above $61 billion, trebling it since the start of the year, and the combined value (April 8) of American Airlines, Hilton and Expedia. One software analyst, Alex Zukin at RBC Capital, described Zoom's returns to the FT as possibly 'the greatest quarter in enterprise software history.'

Yet even as the category definer, Zoom was far from the only star player. Each had its position: one of the platforms, Blue Jeans, emphasised usability: "Ultimately what we're trying to do," said its CCO Stu Araron, "is make video conferencing as comfortable and as casual as your pair of jeans." There was Teams, Webex, Skype (whose loss of category definition stats must surely go down as one of the great business missed opportunities of all time), a confusing medley from Google of Hangouts, Classroom and Meet (billed by programmatic consumer tech radio host KimKomando as offering 'winning security' over Zoom) Jitsi and countless more. EdTech companies like Lingumi, Satchel, Firefly Learning – who run sites for online education – also saw surging demand and revenue, on top of already steep UK investment growth in the sector, 91% between 2018 and 2019. This went for online training too: the UK's Tech Nation's Digital Business Academy saw record uptake in the first two months of the pandemic.

Getting creative - on live video

"It is during our darkest moments that we must focus to see the light," said the prototypical billionaire superyacht owner Aristotle Onassis. In 2020, the dark moments were the all-too-real pandemic and its hundreds of thousands of tragic deaths. And the light was the lowish resolution, 720 pixel width screen resolution of the video conferencing tool.

The chapters that follow are about the dimensions and the implications of the Zoom revolution; not the first word on it, not the last – but a taste of an evolving phenomenon in progress, a real-time exercise in social listening.

Social TV expert Diva Rodriguez considers Entertainment and the rapid-fire revolution in Hollywood that saw major television programmes produced by iPhone and Zoom almost overnight, and production processes instantly changed in ways that may not reverse. Stars that did best on TV were often those – like the Kardashians – who were already used to the self-shooting methodology of the Instagram or YouTube creator. In its attritional war, network TV lost another battle to the digital natives and streamers.

We look also at the Arts – Theatre and Choral Music – and how the development model dramatically changed in moments, at times for the better. Pegram Harrison of Oxford University relates how a choir that set itself the challenge of performing Bach's dramatic and complex St John Passion online reveals as much translatable information about management skills and self-awareness as it does about low-latency telecoms connections and 18th century choral traditions. And it lends practical tips too: set a conductor's guide track, split the audio and video feeds for separate post-production, strip out background sounds, then reassemble.

Michael Diamond of New York University recalls how Shakespeare dashed off *King Lear* and *Macbeth* in a previous lockdown. He reveals how modern playwrights, excluded from

Broadway theatres, used Zoom to create uniquely 'lean-In' moments of eavesdropping intimacy between audience and actors - but found getting it right remains a challenge.

Sports always has a big role in the development of video technology, and a matching prominence this book. A chapter from the British Olympic Association reveals how they used video conferencing and the parallel live chat conversation to make Town Hall meetings inclusive. They re-organised and re-motivated 33 sports teams from Athletics to Showjumping, Canoeing to Cycling, in a dramatic change of plans from "Tokyo 2020", to "Tokyo 2021 everyone is still hoping." They found that organisations may need licences for multiple platforms to have the most appropriate and effective solution for their particular video interactions. And sometimes the best option is two use two videoconference tools at once: a BBC broadcast interview used Skype for the best audio, while recording the presenters on Zoom for mirrored video.

In 2020, eSports boomed, lending their technology of virtual fan populations to real sports with empty stadia, reports digital specialist Alex Fenton of Salford University. He considers 2D video conferencing technology will evolve into 3D video and Virtual Reality (such as in the NBA already). It's a matchday plan that other industries are sure to follow.

Over the 2D video horizon

In business, we learn from PR guru Angie Moxham that video conferencing can provide rich insights into the real lives of people we so far have only known within the dry space of offices. Zoom promotes the empathy and authenticity that are often masked by suits and protocol. But three dimensional non-verbal cues matter – and that's hard to read in the 2D landscape of the Zoom call.

Software isn't just a tool, but actually a determinant of experience and outcomes, as Gordon Fletcher of Salford University explains in a fascinating piece of digital sociology. The complex criteria used by IT departments to buy software often omits the two killer factors – ease of use, and what partner firms are using. The more a customer base is likely to be B2C (say, in education) the more important it is to pick a tool with a free offer – so that the user at the other end of a conversation will be habituated to the software. Meanwhile the 'mosaic war' between video conference providers – from 2 people to 9, to 49 – is adrenalizing function upgrades. But Gordon echoes the point on 2D and its limitations: equal sized faces on a video conference mosaic do not accurately reflect the parallax of a real lecture or meeting, where some participants show more participation or status. Replicating these visual cues of differential roles in a room could be a next step for Zoom and similar tools.

On teaching, David Brook pools his multi-faceted experiences – CMO for major TV channels, Caribbean sports pundit and university lecturer – to find (and confirm in a survey) that multichannel learning actually teaches better. The simultaneous use of conferencing technologies of Blue Jeans and Skype facilitated live Caribbean cricket commentary; and using several layers of communication at once works in the classroom.

And finally - we look at Medicine. Former senior WHO and DFID (as was) official Robin Gorna explains how, at the highest level, global health and international development took to Zoom early as a way of managing vast travel budgets and complex conference organisation. The delicate nuances of international negotiation, the offline 'brush-past' meetings that shift positions, were compromised by the practical barriers to in-person meetings.

End-of-life video

We consider a heart-rending topic not capable of flippant or technology-driven interpretative discussion: the distressing, but necessary communications revolution that has been the arrival of intensive care video conversations between patients and family members amid over 500,000 global deaths by June 2020. On Facetime or Skype links rather than in person, they minimise risk of infection. It's unthinkably acute, but it's real, and just one of the ways Ian Haig describes in which video is metamorphosing in hospitals; an extraordinary change brought about by Covid 19 whose implications will occupy sociologists for years to come.

In looking at hospital life and death, home life, business life and more, the purpose of this book is a simple one: to provoke thought and capture innovation.

We've all been told: "You're On Mute !" For the topics in this book at least, not any more.

About the editor

Dr Alex Connock is Fellow in Management Practice (Marketing) at Oxford University's Said Business School, and Lecturer in Management at St Hugh's College, Oxford. He is also Senior Tutor at the National Film and Television School of the Creative Entrepreneurship Masters programme, and entrepreneur in residence at European Business School INSEAD. His PhD was about video for eCommerce. In his earlier career as a media executive he was six times shortlisted as Entrepreneur of the Year, and Chief Executive of the group that delivered the Teachers TV project to the UK government, one of the earliest at-scale video education projects globally. He is a board member at UNICEF UK and the Halle Orchestra. He specialises in the fast-growing global space where digital marketing and entertainment converge.

Reference

1. https://medium.com/@ezelby/remote-work-market-map-58591966b0c2

Chapter 1

Hollywood: Out of Office

Diva Rodriguez

Summary

When entertainment was most needed in lockdown – it was near impossible to make. Viewing soared but production stopped, leaving producers to re-invent programme making and highly-paid stars to become 'creators.' Video conferencing saturated entertainment - from *Hamilton* on Zoom, to self-shot *Kardashians*, to the no.1 box office hit *Unsubscribe*.

Turning the TV on

Since its 'Golden Age' of the 1950s – ratings the size of national populations, onscreen cigarette advertising, sexist tropes in family viewing - television has been the go-to entertainment destination for billions worldwide.

Indeed, during 2020's COVID-19, viewership still skyrocketed across the globe. People were stuck indoors, literally a captive audience. In the United States, viewing of the big four broadcast networks ABC, NBC, CBS and Fox increased nearly 19 per cent. In the UK, live TV viewing rose 17 per cent from the beginning of coronavirus lockdown, with entertainment shows like *Saturday Night Takeaway*, filmed for the first time without a studio audience, attracting 11m viewers, a record for the ITV show.

The need for TV was high - but there was one problem: the supply chain was stopped in its tracks.

Turning the TV Producers off

Like most sectors globally, TV production was upended by the coronavirus pandemic, and practically overnight.

Television can be watched from the safety and solitude of the homestead, but programme-making involves a *de facto* factory of people on location or a studio set (plus a nice buffet). One day crews filming; the next, they were at home, without air dates or certain resumption. Reality show *Big Brother Canada* Season 8 forecast a worrying future when it was cancelled midseason, 25 days into an 83 days live broadcast. It left a 1.2 million viewership hole in Global TV's schedule.

With productions and live experiences like sports and music halted, what were broadcasters to do ? Could they plug gaps in the schedule with endless re-runs or, worse, would viewers simply turn their TVs off and accelerate the shift to streaming ?

Enter the Digital Creators

Online content creators were already taking advantage of the Stay at Home orders, peacocking their skills as adaptive risk-takers and global distributors to capitalise on the increased demand. The *LADbible* Group launched new and returning original short-form content such as the *Corona Diaries*, a weekly issues-based series. Viral hits *The Gap* and *Agree to Disagree* were commissioned to engage their 50 million followers.

YouTuber James Charles asked his 6 million Twitter followers to tweet their meeting IDs for a surprise "Zoom bombing" to be filmed for his YouTube channel. The result: nearly 3 million video views.

Influencers pulled live audiences of millions with gym classes and isolation tutorials. British reality-star-turned-workout-YouTuber Joe Wicks racked up over 70 million views, as families sought relief. Musicians traded tour busses for selfie sticks to livestream from their backyards. The *One World: Together At Home*, a 6 hour broadcast to raise money for WHO, was produced with artists and production teams working remotely. It went on 26 global networks, plus online platforms, showcased 150 pieces of content and drew 21 million viewers.

Even the big-leaguers of Hollywood got into the digital act - with virtual challenges, singalongs and innovative storytelling. YouTuber Eric Tabach and filmmaker Christian Nilsson wrote, directed and produced *Unsubscribe*, a 29-minute horror film shot completely on Zoom. Craftily taking advantage of a lack of big film releases, the short film topped the S box office, having made $25,488 in ticket sales on the day of release. If the production budget was actually the reported $0, that was the highest ROI in the history of cinema, eclipsing even *Paranormal Activity* and *Blair Witch*.

Actor John Krasinski created the feel-good YouTube series *Some Good News,* featuring monumental moments like cast reunions from *The Office* and *Hamilton*. Eye watering viewing figures of 24 million views and counting were made possible through the use of one key technology: Zoom.

TV pivots to Zoom

With content creation cheaper and devoid of a crew to produce, TV bosses risked losing their leading share of the audience if they didn't act fast. So they did. From writing to filming, everything needed to create a television show had to happen in a virtual environment. Zoom replaced production offices and studio sets.

The writers' room for the 11th season of AMC's *The Walking Dead* had been in session for a few months when the pandemic switched the team to Zoom. Group calls between writers lasted 4-5 hours, averaging less time than a standard day in the office.

CBS's courtroom drama *All Rise* used Zoom to create the season finale, expanding their storylines to include the pandemic. Integrating the current situation into the lives of the characters, the show used Zoom onscreen as a way of mirroring the use of the product by its viewership. The episode was filmed, edited and directed remotely using Zoom, watched by 5 million people.

The producers and cast of NBC comedy *Community* highlighted the practicalities of video conferencing whilst engaging their audience to unite around a good cause. *Community* filmed and aired their Zoom table readthrough for their 5th season on the television brand's YouTube channel in aid of charity. The video amassed 3 million views and raised funds by way of donations to the World Central Kitchen and Frontline Foods.

Mainstream TV Stars become 'Creators'

From Entertainment to Drama, talent was hit a paradigm shift from contributor to creator. But that was an easy win for social media superstars.

The Kardashian family pivoted from influencers to line producers on the 18th season of *Keeping Up With The Kardashians*. Having 550 million Instagram followers and a team to create content, they were forced to self-shoot their television show during the pandemic. iPhones were delivered weekly as the production team sent the director of photography and head of lighting to set up interview rooms at the houses of each of the quarantined Kardashians, leaving directions on the equipment about which buttons to press and how to use the cameras. Producers asked questions over videoconferencing to create the intimate interviews integral to the show's format.

In the UK, ITV aired a brand new 4-part, 15 minute mini-series completely about, and created within, isolation lockdown. The *Isolation Stories*, relied heavily on Zoom.

Sanitised kit - including Samsung Galaxy S10s and S20s, lightweight tripods, a DJI Osmo mobile phone gimbal and a set of LED lights - were dispatched to the cast and their families. Each of the 'filming units' received their own production team including a director, director of photography, first assistant director, remote camera operator, technical and sound supervisor. The teams were able to communicate with the families and each other on Zoom, providing advice on how to set up the shots, adjust the lighting and arrange props. Using screen sharing app ISL Light, technical operators integrated playback with the Zoom app so that productions teams, cast and directors could feedback as the episodes were assembled. The first episode got 3 million viewers and won the 9PM prime time slot.

The BBC revisited playwright Alan Bennett's *Talking Heads* monologues featuring an all-star cast including Jodie Comer, Imelda Staunton and Martin Freeman. Instead of asking the stars to film from home, a skeletal crew filmed at Elstree Studios on the existing sets of continuing dramas forced to stop filming due to COVID-19. The production team minimised contact until the moment of shooting, with the actors and directors rehearsing via Zoom. Make-up artists provided each actor with a sanitised make-up kit and gave them online tutorials about how to apply it; while costume designers went through their wardrobes on video conferences.

So did Zoom save American entertainment ?

In the US, *The Tonight Show* crew filmed the show's last TV episode in their New York City studio on 12th March without an audience, before shutting down for the foreseeable future. The production team met virtually to plan shows with Fallon and his wife in their home, creating *The Tonight Show Starring Jimmy Fallon: At Home Edition.* Fallon filmed the entire show on iPhones. The 10-minute segments ran on YouTube with NBC airing the videos along with rerun material. Echoing the

Keeping Up With the Kardashians team, the *Tonight Show* producers felt Fallon could confidently film from home due to his range of skills and the virtual support from the production team. Fallon interviewed superstar guests using Zoom and experimented with interactivity. The *At Home Edition* version of the show amassed an incredible 300 million views on YouTube. If you were looking for a poster child for the paradigm shift from TV to digital – there it is.

Zoom also became an integral part both in front and behind the cameras with NBC's *Saturday Night Live* team. Writers and actors could create sketches and jokes in team forums, finalising the structure of the show. Self-shot lockdown episodes featured a dedicated Zoom sketch, Zoom-related punchlines, and a 'Weekend Update' news segment created using the team productivity software. The '*SNL At Home*' episodes averaged 6.7 million total viewers, often achieving the primetime programming No. 1 slot on a Saturday night in the United States.

American Idol became the first reality competition show to air a completely remote episode. Producers sanitised and shipped identical boxes of recording equipment to each of the 20 participants including iPhones, tripods, lighting equipment and high-quality microphones. Engineers were sent to the contestants' homes to ensure various Internet connections could handle broadcast TV-calibre production. Producers held Zoom meetings with contestants to help them create the best space to perform, directing the contestants' filming and how they communicate during performances. With the contestants, crews and producers scattered across North America, the show was put together from 25 locations. The virtual finale brought in 7.3 million viewers, hitting a seven-week viewership high.

Even post production took place in a virtual world, with editors often in makeshift suites within spare bedrooms. Due to the cancelling of the *Eurovision Song Contest*, the BBC devised a highlight show entitled, *'Eurovision A-Z'* that mixed live and memorable moments of all time following the world premiere of Eurovision: Shine a Light, live from the Netherlands. The *'Eurovision A-Z'* post-production team used Newtek's NDI platform to stream the Avid output directly to producers. Voice-over record (to picture) session was setup between host Rylan Clark-Neal's home and a dubbing mixer's home studio utilising Source Elements, Zoom and Skype calls and a direct audio feed from Pro Tools. *'Eurovision A-Z'* transmitted to over 1 million UK viewers.

Blockbuster videos

One thing holds evident: Television remains a global source of entertainment, in no small part thanks to video conferencing. Through its adaptive resourcefulness in using Zoom and digital media, television production has managed to weather an ever-evolving and uncertain situation like a pandemic, solidifying its importance within the lives of billions worldwide.

About the contributor

Diva Rodriguez is a digital storytelling specialist with over 10 years' experience in digital marketing and content production, and a full member of BAFTA. She has been shortlisted at numerous awards including the Royal Television Society for excellence in multi-platform social marketing and PR campaigns. She has worked with global organisations such as the U.N., BBC and ITV and her proven record in digital strategy has seen her build monumental social communities. In her spare time, Diva can be found behind her laptop screen penning scripts and sketches.

Chapter 2

Theatre: All the World's a Zoom

Michael Diamond

Summary

Theatre producers have a deep experience of live shows shut down by plague. Early in the 17th century, Shakespeare used pestilential downtime to write some of his greatest works. Matching that is a stretch, because Zoom is like 'really, really bad television'. But has creating and performing plays online during lockdown given us valuable new tools - for art and business?

Creative Lockdown

In 1606, the authorities closed down the London theatre after an outbreak of the bubonic plague. Undaunted, William Shakespeare took advantage of lockdown and rebounded with some of his greatest works – *King Lear*, *Macbeth*, and *Antony and Cleopatra*.

Centuries later, in the Spring of 2020, our own plague to contend with, theatre professionals face existential questions of how we go on: what it means to make theatre when we cannot physically gather in the same place, and how we train and shape future generations? If the lessons of Jacobean drama apply, then theatre might emerge more potent than before.

It is too early to call how productive theatre artists have been during the physical distancing and ban on public gatherings. However, Zoom has not yet been the adrenaline shot one might imagine from such a 'performative' technology. From playwrights, actors, producers and deans of drama schools, a nuanced picture emerges of actors and theatre artists needing to express themselves in a time of collective pain, constrained by a medium many feel ill-suited to this work.

Put another way: *can* you make great theatre on Zoom ?

Zoom stagecraft

Henry Goodman - stage, TV, and film actor - was rehearsing for the lead in Brecht's "The Life of Galileo" at the Chichester Festival Theatre when management decided to close it. From a busy year of filming, teaching and theatre, the frenetic pace hit pause, and Goodman's time shifted to more intimate radio plays. Adeptly swapping .wav files between actors in London and Manchester who never physically met, the BBC director stitched the work together into an artistic whole, with a new hero of this process emerging - the digital audio engineer. Contemplating a Summer of teaching aspiring actors, Goodman reflects that "Our job as actors is to be private in public." The critical question is: "How do you get student-actors to create and work privately, in a Zoom call ?"

James Bundy, Dean of the Yale School of Drama, source of some of the finest actors in American theatre from Henry Winkler ("The Fonz") through Meryl Streep and more recently Liev Schreiber and Lupita Nyong'o, accepts that Zoom is "like really, really bad television." But the silver lining for the students he trains is that it has "taken the pressure off of them to over-energize."

He hears his colleagues say "we have to redefine what theatre is," but for Bundy, Zoom will not be the tool of choice. It's a "dry medium with no feedback loop." The theatre on Zoom he has seen strikes him as a highly degraded experience: "No one has figured out what the events are, and lots of energy is going into which quadrant of the screen you are occupying."

But an intimate theatre piece premiered by New York's Public Theater - *What Do We Need To Talk About? Conversations on Zoom* – suggests there is hope. From a company typically fielding 3,000 seats across six stages, and in a sign of the power of digital distribution and the economic disruption of pandemic, Richard Nelson's new play premiered free on YouTube. Nelson revisits his long-serving characters – members of the Apple family - from the "Rhinebeck Panorama" series. This time on Zoom, their conversations woven into a beautifully curated, moving tale of loss, love and the power of story-telling in a pandemic.

Fittingly, Barbara, eldest of the clan, folds in the *Decameron*, 14th-century Florentine stories told as Boccaccio's protagonists sheltered in place outside their plague-ridden city. Of earlier plays in the Rhinebeck cycle, Nelson has been quoted in interviews suggesting that he wanted characters on stage who "had no interest in making any points but simply wanted to talk to each other and listen to each other … to be, not to show." As observers in this new private drama on Zoom, the 'audience' leans in to whispers between siblings. We share the anxieties of Jane, the youngest sibling, and the discomfort of her partner Tim, quarantined in a separate room recovering from the Coronavirus. We hang on painful memories evoked as characters pause to remember the death of a sibling's child.

Exploring a new medium

Yale's Dean Bundy suggests "there is some room in the medium to take advantage of the medium." But in less well-trained hands he sees student-actors "composing the space" when rehearsing. This is "valuable but highly performative, and not necessarily revelatory of the kind of human experience that we want from actors whether they are on film or on stage … They can create certain kinds of atmosphere, but they would never do that, if they weren't on Zoom." These strained reactions are partly innate to the software, which privileges

24

one conversation, one person speaking, and defaults the fixed camera to a static view.

The BBC's series *Staged* tackles many of these issues head on, with actors David Tennant and Michael Sheen delivering what the critic and comedian Viv Groskop described, on Radio 4's *Front Row*, as "the purest 15 minutes of joy" after watching the first episode. Together these friends explore the limits of "creating great drama and great comedy using these really limited techniques of self-filming and Zoom," Groskop observed -- creating "something new and original, that references this moment without being trapped in it."

Notable student productions may also be breaking the mould, like Bard College's Zoom production of Caryl Churchill's play *Mad Forest,* which the organizers saw "not as a replacement for live theatre, but as another way performance can happen and a story can be told." The program notes published online draw parallels to the historical revolution in which the play is grounded: "while the webcam flattens visual information, Churchill's material takes on new dimensions in this space. It has been noted that television and the amateur camcorder fundamentally shaped the message of the Romanian Revolution."

Happy accidents

Shay Wafer, involved with both 651 Arts in New York and WACO (Where Art Can Occur) in Los Angeles, celebrates Zoom for expanding the digital creation and distribution of art, remarking that some theatres see opportunity in this technology shift. Much larger audiences can be assembled online for performing art that would previously enjoy more limited local exposure.

Bill Langan, Assistant Professor of Acting at Oklahoma City University, worries though about a loss of 'happy accidents' — "one of the things we rehearse for ... things that happen because an actor made a surprising choice and it informs the other actor's choice, and we are off the to the races, we are off in a different direction than we were yesterday .. and this can't happen on Zoom."

One such recent accident however, might not have been welcome. In California the ZoomTheater is creating live shared events with audience feedback. Introducing a production of Anna Ziegler's *Actually* the director reflected that Zoom was "imperfect for this purpose ... not designed for live theater" but encouraged us to "think about it as the early days of cinema [and television] with this extraordinary technology kinescope ... exploring the potential [of the medium]." The audience was guided through the protocol of mute on or off, rules about eating and drinking, dogs barking and babies crying. But nothing could prepare us for the ZoomBombing of the production on the night I attended, with harsh racist language rapidly filling up the chat box and racial epithets barked from a few "audience" members. The Stage Manager, now with digital jurisdiction, had to eject them from the 'theatre.'

Robert Pinsky's essay *Responsibilities of the Poet* suggested that "an artist needs, not so much an audience, as to feel a need to answer, a promise to respond." For Goodman, as an actor, this "promise to respond" is a critical fulcrum, as he reflects on theatre in an age of Zoom: "Are we just interpreters of writers of words, or are we instigators with urgent needs of our own, which are not only dependent on writers ... what bit of us can function without an audience?

Whether the theatre itself can function without an audience is an existential issue, but Zoom is clearly changing the game. As Emily Mann of Princeton's McCarter Theater said, in a Facebook Livestream event: "I think that when we come out of this, we're going to be looking at making theater in different ways ... we're going to have to strip down to the essentials, and the essence, of what theater is all about." Oskar Eustis, Artistic Director of the Public Theater, home for Nelson's new play, agrees, added to Mann's comments a grounded sense of what these times of economic hardship also demand: "we have to be very clear about why we're necessary. There's going to be a huge demand on us to prove that we are an important part of the recovery from this because the demand for resources is going to be everywhere."

About the contributor

Michael Diamond is the Academic Director of the Integrated Marketing and Communications department at NYU's School of Professional Studies, where he is a Clinical Assistant Professor in the Division of Programs in Business. Diamond also teaches at the Yale School of Drama, and had an early career managing and fundraising for theatres in the UK and the USA. A graduate of Oxford, Yale, and the London Business School; prior to his academic role Diamond worked at Time Warner, leading initiatives in strategy, business development and marketing. Michael is proud to acknowledge that Henry Goodman is his uncle.

Chapter 3

Sport: The Zoom Olympics

Scott Field

Summary

When the Tokyo Olympics was rescheduled from 2020 to 2021 at short notice, the British Olympic Association had to re-plan performance peak and logistics for athletes from 33 sports – without being able to meet any of them. They used every major video tool to do it.

Tokyo 2020: postponed

In mid-March 2020, a staff delegation from the British Olympic Association – home to Team GB –stepped off a plane back from Tokyo, having spent a final week on the ground in the Olympic city, testing and stressing Team GB's Games-time plans. They weren't over confident – but everything seemed to be falling into place for the 33 sports teams they would be sending from the UK to compete in three months' time.

It didn't turn out that way. Little did they know that within days, Japan would go into lockdown, and the Tokyo 2020 Olympic Games would be postponed and shifted a whole year to 2021.

The planning of the Games is, in the words of IOC President Thomas Bach, like piecing together the world's biggest jigsaw puzzle. An Olympic Village hosting near 11,000 athletes, over 20,000 members of accredited media, and 40 concurrent world championships running side-by-side, make for the greatest and most complex show on earth.

For Team GB, the postponement meant the renegotiation of every contract with every supplier, staff member, volunteer and commercial partner, both domestically and in Tokyo.

It needed to be done quickly, sensitively and with absolute clarity to ensure our meticulous planning for the original event wasn't compromised for 2021, and to protect every athletes' chance of performing to their very best. And it all had to be done from home.

The solution to getting that done was face-to-face video-conferencing, whose steep acceptance was vital. So what did we learn?

The ancient sport of video conferencing

When it comes to trust, human instinct tells us to question most things and people. In the office environment there is none more questioned than the home worker. Such outrageous requests as to want to work from home for a single day a week have often, in the past, been met with an air of suspicion and wariness.

Since COVID, we have seen a forced, yet seismic shift in attitudes to working from home, driven primarily by the rapid uptake of video technology to keep businesses functioning.

But it's the exploration of the true capabilities of these technologies that has been the real game-changer, for we have long-since known of their existence, and in many cases have already been using them. Just not very well.

Just as we tended to kid ourselves that the notion of 'flexible working' was a thing before COVID, we have awoken to find that having a conferencing facility in the office – often in a vacuous meeting room, with software tethered to a PC locked in a back cupboard, that only can only be operated by IT personnel – was as much the result of the vanity project than it was of any actual commitment to implement smarter ways of working.

Despite having had open access to Skype, FaceTime and other bespoke platforms to be able to manage conference and video calls in the past, few people had ever been shown or encouraged to explore their true functionality. Using audio-only options led to poor etiquette and often wasted opportunities, with overly verbose conversations between those present in office meeting rooms leaving the lone home worker rendered mute for entire sessions.

Complex operating systems in meeting rooms, reliant on hard-wired microphone or speaker connections have been quickly forgotten with the adoption of simple plug and play applications that require no more than the user to look into their laptops' web cam and find the unmute button.

Online video decathlon

With the Olympic rescheduling – and a million other coronavirus-driven disruptions world wide - all of sudden, the simplicity of the home office/bedroom/playroom/kitchen has overtaken what now look like overly bureaucratic office processes and policies to deliver more seamless experiences for the end user.

Now we're all home workers it seems that everything has changed. We have quickly become adept at navigating the myriad options available to us – in a single working week recently this author used FaceTime, Skype for Business, Zoom, Microsoft Teams, Google Hangouts and Cisco Webex to speak to a range of internal and external colleagues.

There is a rapid normalisation of video calling across the business landscape and a ready acceptance that for domestic organisations – just has been the case for multi-nationals for many years already – that you can operate normally, and at a high capacity, very well from home or out of the office.

With dozens of hours of commuting time back in the bank already, you can now add several more hours of time saved through more focussed meetings, with often clearer objectives and outcomes.

Flexibility in approaching the technology is important as we navigate different in-platform features, security and, for organisations like the BOA, even the ability to turn these interactions into broadcast quality content for our external digital channels. On that basis, it may well be that organisations require licences for multiple platforms for the foreseeable future to ensure they are able to choose the most appropriate and effective solution for their particular video interactions. A recent BBC broadcast interview saw us calling in via Skype to provide the best broadcast quality audio for radio, whilst simultaneously joining the presenters of the show on Zoom to create a mirrored video recording.

Virtual Team Spirit

There are further advantages in the ability to show multiple users on-screen at any one time, especially when facilitating large numbers of participants, but also in the case of wanting to build and maintain team culture and spirit.

This is a challenge for every organisation that has experienced the overnight and sudden shift from office working to home working, especially when you exist in a business that is built on team ethos and culture, such as a sports team like Team GB is. Necessity is the mother of invention, and the development of a virtual social culture to supplement the daily formal business commitments has shown that culture can still flourish.

Audio quality remains an issue for many too, but there is already a growing etiquette amongst video conference users, which has led to both increased sound quality for listeners and opportunity for everyone to speak and, more importantly, be heard in team meetings.

That said, the growing integration of chat facilities and users' willingness to engage in live text whilst taking in the content remains a challenge, especially when faced with large groups. Many virtual 'town halls' may still feel one-directional in respect of content delivery with little interaction afforded unless expressly requested or prompted. The need for interactivity on chat functions – through questions and debate – is important if we are to really engage the workforce in the way we might through traditional face-to-face interactions.

New considerations for occupational health will soon come to the fore, as the long-term outlook for flexible working requires suitable workstations and environments from a health and welfare perspective, as well as the challenge for businesses to support the funding of such hardware.

At the BOA all users have their own headsets and landline phone facilities have long since been delivered through a desktop and mobile solution which requires no physical handset in the office and is therefore 24/7 accessible.

All of this is, of course, predicated on digital accessibility and employers can ill afford to leave workers digitally deprived either through lack of access to high-speed wireless connections or basic lack of training and skills. This will be balanced in time, by the reopening of office environments and flexibility being afforded not to those that want to work at home, but for those that want to brave the commute and be back in HQ.

The future has changed. For those who work for Team GB, it is unlikely they will ever go back to a rigid 5-day-a-week, 9am-to-5pm pattern. Video calling and conferencing has been normalised. The technology has held-up and the response from the workforce has been overwhelmingly positive.

Importantly though, through all of this fast-track learning, we're also building trust, or at least eroding the suspicion of the work-from-home opportunity, as visibility of colleagues becomes tangible proof that there really was another way to work all along.

About the contributor

Scott Field is Director of Marketing and Communications, British Olympic Association

Chapter 4

Education: Lessons from Live TV Sports

David Brook

Summary

Commentating remotely on Caribbean cricket by using two different video conference tools simultaneously inspired a multichannel approach in universities. Opening video and chat as completely different streams gave an open feedback loop that's a hit with students.

Multichannel cricket

We were having a debate on the Sportsmax Zone's Ultimate Test XI, about whether South Africa's now-retired and prolific batsman Jacques Kallis deserved a place. (You might think it's niche – but not in the Caribbean, where in many places cricket is still a national obsession.)

"We talk about players that empty bars, and he wasn't one of them," I said to my co-hosts on the Guerilla Cricket show. "He may have scored 45 centuries, but I can't remember a single one of them."

Sportsmax is the leading pay-TV sports network in the Caribbean. Broadcasting from its headquarters in Kingston, Jamaica, producer Ricardo Chambers says that its adoption of videoconferencing is pivotal:

"It means we no longer have to depend on persons coming in to the studio as they can speak to us from their desk or room or location. They can see us and we can see them."

Most interesting, is that they use two video conference channels at the same time: Skype and Bluejeans. One for the audio, and one for the video.

You might think of Bluejeans as classic leisure-wear - rather than a high-spec video conferencing platform. But around the world, the adoption of this nine-year-old software (now owned by Verizon) is having a big impact - on how sports broadcasters present live sporting debate and commentary.

Bluejeans is an interoperable cloud-based video conferencing service, similar to Zoom and Teams, meaning that it connects participants across a wide range of devices and conferencing platforms. Before the Covid-19 crisis, Skype was the go-to application for conducting interviews remotely - but only with one person at any given time. Now the multi-player Bluejeans platform (like Zoom) give broadcasters a lot more options. The TV studio had been seen as sacrosanct, the only place to have discussions and interactions. Not any more.

"As a company, we have been using Bluejeans for meetings well before the crisis" says Chambers, "but now we are seeing there are many more efficient ways to get things done. We have been given a chance to slow down and figure it all out."

Similarly, Sky Sports in the UK are enjoying success with their 'Watch-along' series, where recent cricket archive is given a new lease of life by replaying famous matches in the company of the main protagonists - who contribute via Zoom. Audiences were able to enjoy the 2019 World Cup Final and the Ashes Headingley Test alongside players Ben Stokes and Jo Root, who were in vision from their homes giving insight into their thought processes at the time.

The 2020 Test series between England and West Indies will be played behind closed doors in a bio-secure environment. And again disruptors will be providing coverage on the match via a network of commentators broadcasting off-tube … via Zoom.

Multi-channel zooms into the classroom

There are similar dynamics in college education. The technology may have been around for a while, but it's the lockdown and the pandemic that have forced broadcasters to re-evaluate how they use their resources.

The same process of re-evaluation is having a dramatic impact on how we work, teach and study. Home working during the pandemic had had a positive effect on productivity, according to 54% of respondents in a recent survey of professionals by Linked-in and USA Today. The reasons for this: time saved from commuting (71%), fewer distractions from co-workers (61%) and fewer meetings (39%). Anecdotally, many would agree Zoom meetings encourage more junior members to contribute, as traditional office hierarchies are less pronounced on the flat and therefore democratised mosaic (although it's only a matter of time before some platform restores office power grids by introducing a parallax effect.)

The adoption of video-based real time teaching and lecturing provides the all-important face to face interaction and is helping to re-define online learning away from its roots in pre-recorded webinars and self-assessment. The combination of the live lecture, interactive conversation with students on video, matched with a second feed of text discussion (chat) sometimes on allied but not overlapping topics, and Q&A which can be systematically checked off by the teacher, simultaneously informalises and structures the learning in ways not possible on the singular plane of a classroom lecture. 'Virtual face to face learning' is a more accurate description.

Educators can no longer rely on the 'captive audience' of the lecture theatre. They have learned from live TV (and other media) that audience engagement is key when it comes to screen-based interaction. That engagement requires video, audio, live chat and break-out rooms to keep learners from switching off mentally or physically. This represents a paradigm shift in education provision.

Research: do students like online learning ?

From our own polling conducted during lockdown amongst over 500 students at Oxford Business College,[1] the move towards screen based educational provision has been welcomed by students – and substantially. 64% of respondents would like to see all teaching delivered in this way, even after lock-down has eased. And as many as 83% of respondents would like to see some form of virtual teaching remain. When asked why they liked virtual teaching, 65% agreed it was saving them time and money, whilst 27% agreed it was safer for their health.

However, the primacy of bricks and mortar education is likely to remain for a good deal longer. In Scotland, First Minister, Nicola Sturgeon said proposals for "blended learning" will not continue "a single second longer than is absolutely necessary" (STV News, 15 June 2020). "We want young people to be back having face-to-face teaching for 100% of the school week as soon as it is feasible."

But perhaps educators should watch a little more Caribbean cricket. Stop thinking of video conferencing as a stop-gap, learn from the adoption of these platforms by sports broadcasters and news providers, consider the benefits of multi-channel two-way communication, Blue Jeans and Skype, and reconsider what "face-to-face" really means in this digital age.

About the contributor

David Brook lectures in Business Strategy, Marketing and Digital Media at Oxford Business College and Oxford Media and Business School. He will be providing cricket commentary and reports on the forthcoming England vs West Indies series for Guerrilla Cricket in the UK and Sportsmax TV across the Caribbean. He was previously Director of Strategy, Marketing and Sport for Channel 4 Television.

Reference

1. In-house research via Zoom polling, May 2020, Oxford Business College.

Chapter 5

Hospitals: Video Technology in the NHS During the Covid-19 Pandemic

Ian Haig

Summary

Coronavirus has driven dramatic changes in how hospitals use live video. The need to control the spread of infection meant that visitors were no longer allowed inside. Video became a vital link between patients and their families, offering solace at the most distressing of times, including in the final days of life. Patients without Covid-19 saw their regular face-to-face appointments changed to a video call with their doctor as habits, formed over hundreds of years of medicine, were transformed overnight.

Coronavirus on Camera

Intensive care units across the world had to adapt in early 2020 to a new threat to everyone's health – coronavirus. Infected patients deteriorated so quickly at home that they had to be rushed to hospital. For many, this would be the last time they saw their loved ones in person. The risk of transmission in our hospitals meant that, early on in the pandemic, visiting restrictions were put in place, preventing the usual bedside vigil from friends and family. The fear and agony of that experience for many was unimaginable.

Covid patients arriving in hospital were surrounded by brilliant staff, but to keep everyone safe protective equipment was essential – masks, gloves, aprons. For the people behind the masks (doctors, nurses, and therapists) it was an uncomfortable experience – hot, sweaty, and painful on the skin. These committed professionals desperately wanted to reach out to their patients, to smile and comfort them, but the coronavirus made all of this so much harder.

So when Neil Anderson saw a coronavirus patient speak to his family via a video call for the very time, on an iPad donated by Barts Charity, it was a revelatory experience.

"I've never seen anyone that happy", says Neil, a critical care matron at Britain's oldest hospital, St Bartholomew's in London.

"The joy on his face was incredible… his whole face lit up like I've never seen before.

"Nothing can match being there in person, but that was the moment when I realised that video was the next best thing."

Critical care is reserved for the most acutely ill patients and managing concerned family and friends is an important part of the job.

"It's often the families that need more support than the patients," says Neil.

Staff struggled to be heard when they spoke to relatives on the phone, their voices muffled through the mask they were wearing to keep them Covid-free. Often they'd be misheard or misinterpreted, exacerbating the frustration for the person on the other end of the life.

A solution was needed in the form of a 'Covid enquiry hub' – a 24/7 hotline, staffed by a team of nurses, doctors and administrators, based outside of intensive care. The hub staff would be present at shift handovers and at team meetings, picking up intel which they would later relay to families desperately seeking news.

Preparation for live video in ICU

Having a family member admitted to critical care is a distressing experience. Not being able to visit only compounds this anxiety.

"Relatives want to see how they look and how they are", says Leisa Griffiths, a critical care nurse who spent time working in the enquiry hub at St Bartholomew's.

"That's why, for patients who are awake, orientated to where they are, and who can communicate, we ask for consent to set up a video call."

Preparing the family for what to expect is key, says Leisa.

"The patient may look different after weeks in critical care. They may have lost weight and changed in appearance.

"We also ensure the patient knows everyone on the call, so that other people are not peering in, just in case this is upsetting for our patient."

"We request that no pictures are taken during the call. We also have to warn relatives not to be too disappointed if the patient isn't feeling up to it on any day."

When everything aligns, Leisa says, video is a priceless tool.

"I begin the call, say hello and show that I'm in my protective equipment. Then I turn the tablet towards the patient, while I'm standing behind it. You suddenly see the patient's face light up as they recognise their family.

"It's a very special moment."

After days, weeks and possibly months apart, what is the first thing that patients and their families talk about?

"It's funny", says Leisa.

"They talk about lots of things that you'd think were perfectly mundane...who's looking after the flowers? Who's doing the cooking?

"One relative said she was off to have her fish and chips and the patient replied: 'it's all right for you. I can't eat!'"

Humour is important, says Leisa.

"One patient pressed the disconnect button during a call.

"I thought, 'he's done that accidentally', so I rang back but he pressed the button again.

"I called his relative to explain and she laughed... 'that's okay, I know he's getting back to his normal cheeky self now!'

"It's great getting to really know your patients and their relatives, and what makes them tick.

"Often families send us photos and letters, which we laminate and share with our patients.

"We have to do everything we can."

Hospital clinics on video

The need to keep people out of hospitals at the peak of the pandemic meant that, overnight, millions of face-to-face appointments were changed to telephone and video appointments – something the NHS has been trying to achieve for more than a decade.

For St Bartholomew's consultant Anthony Bastin, this has meant his clinic, for patients recently discharged from critical care, could continue.

"Video was something we were meaning to get going with" says Anthony.

"COVID accelerated this to happen overnight."

Anthony, who uses a platform called Attend Anywhere to host his video clinics, says one significant benefit is the time and money saved by patients and their families in not having to travel to hospital.

Anthony's patients often have symptoms of post-traumatic stress disorder and depression, social interaction difficulties and many other issues after such a traumatic experience.

He says the calls can provide a valuable insight into the patient's lifestyle.

"Where patients show you, it can be interesting to see the environment where they live as this might offer suggestions for how to improve their health and recovery.

"On the other hand, the quality of the video call really depends on the internet connection. It works much better if a patient uses a computer instead of a phone. There can also be audio issues, which can make it a struggle.

"Although you lose the benefit of human interaction, you can still develop a rapport with the patient, such that it's not far off the real thing.

"Clearly you can't examine them on camera, and it's tricky to assess their walking, but it's working very well."

Death and video

For the vast majority of patients, hospitals have been able to accommodate at least one visitor at the end of life so that no patient dies alone. But is there a place for video at one of life's most precious and personal moments?

This is especially difficult on intensive care wards, where patients often deteriorate so quickly that it's impossible to get consent.

"The patient can't let us know it's okay", says Neil.

"If the patient can't communicate, this can be upsetting for the family.

"There's also the risk that the call might be filmed."

The pandemic has highlighted the issue of friends and relatives being abroad and unable to return in time.

In these circumstances, and where a patient is awake and well enough to communicate, a video call can take place.

Lasting legacy

In three short months the NHS has been transformed in a way that places video at the heart of everyday life, from team meetings to training sessions, video visits and consultations.

At St Bartholomew's Hospital, this has been made possible so quickly thanks to generous donations and the adaptability of our truly amazing staff.

"There was a look of fear on everyone's faces before the first patient with coronavirus came," says Neil.

"When they kept arriving, day after day, teams realised what they needed to do, and got on with the job brilliantly."

Video has been central to this success.

About the contributor

Ian Haig graduated with a masters in Creative Business for Entrepreneurs and Executives at the National Film and Television School (NFTS) in February 2020. His creative background is in the theatrical world, as Producer at Eyebrow Productions and Showtime Challenge: 48-Hour Musicals performing fully-staged musicals at major West End theatre (including the London Palladium) after just 48 hours of intense rehearsals – all in aid of charity. Ian has launched a production company called Springboard Pictures which is concentrating on healthcare projects following his 18 years' experience as Manager and Operations Director at world-leading specialist NHS hospitals. Ian is also proud to play bass trombone for the London Gay Big Band. Ian returned to his role as Director of Operations for the COVID19 response with the amazing team at St Bartholomew's Hospital.

Chapter 6

Global Health: An Early Case of Zoom

Robin Gorn.

Summary

The global health community – valued in 2019 at a vast $11.
trillion – caught the Zoom bug long before Covid. But the nev
virus made it run major global health conferences, and conduc
tricky "political" negotiations online – with results that week
earlier had seemed impossible.

Video conferencing

Awkwardly, the first time I forgot the mute button was on
call with thirty public health and human rights experts, fror
five continents.

It was mid-May 2020 and we were exchanging tips on how t
make sure that gender equity and human rights are at the hea
of US$14 billion that will be spent in the poorest parts of th
world. Countries were busy scaling up services financed by th
Global Fund to Fight AIDS, TB and Malaria (Global Func
while dealing with the double whammy of escalating COVID
19 epidemics.

And I was busy preparing fresh OJ in a noisy electric juice
When I noticed the mic had been on, I was red-faced not onl
because it was obvious that I wasn't paying full attention, b
there was no way I could play the Zoom novice card.

In the international development and global health communities, we know Zoom. For many years I've been happily making breakfast or dinner with some lengthy international call or webinar as background entertainment. When it came to Zoom – a bug that, by mid 2020, would be in the lives of 300m daily users - this was one epidemic where the global health community was far ahead of the curve.

I've worked in global health for over three decades, racking up serious airmiles along the way (and equally serious environmental angst alongside.)

In early 2017, I was responsible for the start up of a new global women's rights movement - *SheDecides*. Within a few months my Dutch co-Lead and I realised there was little point in her flying back and forth from Leiden every week, or for me to struggle on commuter trains from Brighton to London.

We settled on video conferencing.

Global health on a Zoom lens

Most of our daily interactions had us glued to our screens on Blue Jeans (an upmarket alternative to Zoom), noticing that we could be just as effective in different rooms. Our job was to crowd in commitment from politicians and activists in countries across Africa, Asia and Latin America, and to keep brave European and Canadian leaders on board.

We were busy plotting novel campaigns, and our first hire was a talented young communications manager living in Melbourne, Australia. Soon she was joined by a seasoned activist in Ireland, a campaigner in Washington DC, and a young leader from Cameroon.

This was the quintessential, globally distributed workforce.

Eventually we brought all of them to London for chunks of time, but much of our team work was managed through Blue Jeans videocalls and making Slack work for us.

As the movement grew we were in daily contact with activist and movement builders in India, Kenya, Malawi, th Philippines, South Africa and Uganda, as well as Unite Nations (UN) officials in New York and Geneva. Even wit my (then) energetic passion for travel, I could never hav helped the movement thrive without video conferencing.

For almost a decade I've stayed up-to-date on scientific an policy developments on women's health and HIV by staring a my laptop, and even listening to global updates walking alon Brighton beach.

Before this pandemic erupted I joined at least one call ever week: UNICEF webinars on adolescence and HIV, regula strategy updates with Global Fund colleagues, and monthl sessions with the Every Woman Every Child (EWEC Advocacy team. Our EWEC call in May 2019 updated ove 100 globally dispersed colleagues about campaigns to en Maternal Deaths and Obstetric Fistula, to applaud Midwive and to prepare for World No Tobacco Day. Most of the ca was busy discussing the crush of side events scheduled for th 72nd World Health Assembly (WHA) – the annual gatherin of Health Ministers that governs the work of the World Healt Organisation (WHO) - my previous employer and now a trul household name.

In February 2020 I heard the UN Secretary General on th radio. He was urging Ministers *not* to fly to New York fc CSW (Commission on the Status of Women), the annua gathering to review progress on important resolutions taken 2 years ago.

That was the day I knew that this pandemic was out of control

All member states have senior diplomatic representatives the UN, so in theory there is no need for any travel to happe Yet every year Ministers and teams of officials flock in fro all countries: one big fiesta for each of the specialist agencie culminating in the General Assembly (GA) jamboree th brings together Presidents and Prime Ministers and creates gridlock in the streets of New York for a week or two every September.

Tough politics on the small screen

I've attended tens of these UN GAs, WHAs, CSWs and the like. It was unthinkable that they could happen on-line. But in 2020 they all have; no-one expects world leaders to fly in to New York in September.

Since I no longer work for a government or the UN, I'm not privy to how these meetings went, but I'll bet they are nowhere near as politically savvy as they should be, and that advances many have battled for over the years will have slowed dramatically. It's the informal corridor chat where political trading happens – and then it plays out on the floor of the UN or at whatever Board meeting all those leaders have flown in to attend.

Where Zoom gets tricky is when you need to make tough "political" decisions.

When I wasn't noisily juicing my oranges, the *Global Fund Technical Review Pane*l did great work advising how to spend US$ billions. Still the meetings took almost twice as long as if we'd been (as normal) in lockdown in a Geneva hotel rather than in our individual lockdowns, with Asian and Australian experts offering insights and judgments from kitchen tables at 3am.

It took longer in part due to frequently wobbly internet connection for the most distant colleagues, especially in Africa. Less fix-able is the fact that – even with fabulous break-out Zoom rooms - you just couldn't grab a corridor chat or coffee to worry through some knotty problem and small twist in interpretation, or nip over to the WHO building to ask experts to confirm details of recent technical guidance.

Of course we scrolled through written guidance and information, asked Prof Google, sent pleading WhatsApp messages to colleagues, but the Global Health world function – like any other sector - on personal connections and trust. It' not so easy to build that through Zoom.

Missing the 'brush past'

What is special about all of those various UN acronyms, and the Board meetings of Global Health bodies like the Global Fund, is that much of what happens is based on forging personal connections with people from diverse cultures whose written positions, or professional personae, may seem to be at odds with your own. I have been part of massive shifts in global policy (usually on politically hot topics like sexual rights) smoothed by hastily arranged small dinners, or smartly engineered one-to-one chats at cocktail parties in the margins of the formal meetings. In diplo talk these are known as "brush pasts" and – even for these apparently trivial moments - small civil servants prepare their Ministers with Talking Points (TPs) listing bullet points under the headings "Lines to take" and "To raised". It's a well-established trick of the diplomatic trade and global health diplomacy is no different. It's part of why people go to massive professional gatherings like the (20,000 person) International AIDS Conferences that I was once responsible for.

My successor is now busy with the terrifying task of shifting the next one, originally slated for San Francisco/Oakland in July 2020, into a virtual event, with a side meeting on COVID 19 science (since so many AIDS professionals have been repurposed for a new pandemic). He has to bring it on line and still make enough money to run the organisation. It's not obvious.

Formal scientific papers and workshops can easily be presented on Zoom – and there's been a global upswing in appetite for on-line learning. But much of that appetite is driven by freebies (I'm currently digging into my second on-line course – happily learning about film making). They've sliced the fee in half – but will anyone turn up? And how will on-line schmooze work?

The appeal of these massive conferences, and their business model, turns on the rare opportunity to get thousands of like-minded people in the same place for a week. Like trade fairs, global health gatherings are designed to build solidarity, allowing focused, quiet chats and connections that never happen otherwise.

I'm hopeful that as we move to the next stage of living with COVID-19 (no, we will never be "post-pandemic") we become much more judicious and thoughtful about when we meet physically. Our domestic stability, physical and mental health will benefit just as much as the planet from reducing flights to the truly essential.

And we'll hold on to amazing Zoom innovations. This article was edited at my daily "Writers Hour" – a fixed slot where I start the day writing with over 100 people. *The London Writers Salon* organise it, and writers dial in from all over Europe, India and even Australia. We all notice that we get far more done by being together, writing silently with a backdrop patchwork of others staring at words emerging on the screen. It's a brilliant event – we chat for 5 minutes at start and end, and somehow that provides the push and motivation we need. Something I would never get just allocating an hour for myself – or hopping on public transport, losing three hours of my day to do it in person.

About the contributor

Robin Gorna is a global feminist who has led international and local campaigns and organisations since 1986, including

SheDecides, the Partnership for Maternal, Newborn and Child Health (PMNCH) at WHO, and the International AIDS Society. She set up DFID's global AIDS Team in 2003 and spent four years in South Africa leading the UK's HIV and health programmes in the region. She is now working on a memoir of a life lived between two plagues. See https://robingorna.com/

Chapter 7

Technology: How Long until 3D Zoom?

Dr Alex Fenton

Summary

The global lockdown not only adrenalised our conversion to online video communication – but also our move towards immersive video technology. VR and AR have leaped across a gap in our consumption of sport and entertainment. Now there may now be a moment in the 2020s when we move from 2D video to virtual, 3D interaction.

Sports become lockdown eSports

The global lockdown caused by COVID-19 saw events just stop - from major sporting occasions, to tours, festivals and business conventions worldwide. Annual conferences were postponed until 2021, or just cancelled completely. Other events took to video software such as Zoom, Teams or Google Meet and many more.

Switching meetings to Zoom was revolutionary for some in education, sports and entertainment. But in other fields, it was 'business as usual' – because of the dramatic pace of change that was already underway.

Take sports coverage – the 'live' element of which has always been pivotal, driving innovation in live broadcasting from black and white TV to colour, to HD and 4k, satellite, cable and livestream. Heavily consumed through TV and social video, live sport was stopped in its tracks by the pandemic in almost all major leagues globally.

Starved of the content that feeds them, sports clubs innovated fast. There was an accelerated shift from 'real sport' to esport live video.

With real football postponed, UK soccer club Leyton Orient organised a massive 128-team #UltimateQuaranTeam FIFA2020 competition. Famous players competed representing their teams to create a digital FIFA soccer festival, with the live games streamed on Twitch. Clubs, players and fans posted on social media like on a real match.

Globally, virtual competitions were created to fill the sporting gaps. New fans (and gamblers) who had never before watched esports were now following online in events powered by video streamers like Twitch, which in March 2020 saw a 23% increase in usage. This may not sound like a huge rise, until you consider that this is 1.2 billion hours of video streamed in March, enough to play each US citizen the World Cup Final twice. Steam, a video game distribution platform, also hit all time record figures with 24.5 million users on 12 April alone.

It's not just esports too: sports have become an opportunity for a giant multi-level videoconference.

Teams such as AGF Aarhus in the Danish Super league created a virtual grandstand for fans to watch their live game on Zoom. The match was behind closed doors - but allowed fans to not only watch the match, but themselves be broadcast at the match on giant screens. Project co-ordinator Soren Carlsen said, "They will hear it, they will feel it and see it on this big screen alongside the pitch. It's an opportunity for them to get a sense of the community that they experience around football because it's also about meeting people at the stadium (Dean, 2020).

It is this sense of immersive interactivity and community which is so critical to the success of interactive video - in sport, education or entertainment.

Whilst the global pandemic forced forward some innovative video solutions, it also begged the question what next? It presented opportunities to accelerate digital transformation and event integration with data and digital systems, creating even more pleasing immersive video experiences that surpass the live experience 'reality'.

Accelerating into Events 4.0

So what's next?

A sporting and events calendar where the interaction and transmission are happening at multiple levels all at once – both real and virtual.

The suffix 4.0 derives from 'industry 4.0', relating to data exchange in manufacturing, technologies and processes (Gerard et al., 2020). The paper 'Recognizing events 4.0: the digital maturity of events' explores the role of technology and systems in modern events in order to present a definition of 'Events 4.0'; "events that are frequently iterating, digitally managed, fully integrated in their data and digital systems with dedicated technology engagement." Furthermore, they: "optimize communication at all levels in order to inform other sections of the delivery operation to maximise marketing opportunities and enhance the experience of attendees".

We can draw a fictional future example here from Flash Fiction or Micro Science Fiction Prototyping *(μSFP)* as follows:

"Zeke polished the lenses on his holo-sunglasses ready for Glastonbury 2029 where Beyoncé and Elvis are duetting live on the main stage in his garden. His friends appeared one by one with pizza and drinks at the front of the stage. Some of them are real."

It could be argued that current audio and video Zoom calls ar attempting to substitute the face-to-face experience with 'digital twin.' This is a poor cousin perhaps of face to fac experience, with so much lost in social and non-verba communication, but there are myriad of opportunities t advance beyond the video call to more immersive digita systems, and use of data as in the above definition of event 4.0 and our example of Glastonbury 2029. One opportunity fc immersive video: Virtual Reality (VR).

Reality hurts: let's have some VR

The pandemic came a little too early in the life of VR adoptio for an exponential uplift in usage as we saw with Zoom – bu that steep rise is coming.

VR has a long history and what seems like a very slov evolution, but it is developing fast in the 2020s. Headsets suc as PlayStation VR and Oculus continue to grow, and sale spiked further during lockdown.

Facebook have invested heavily in VR headsets with the purchase and development of Oculus Rift, announcing tha 80% year-over-year quarterly growth of "Other" revenu [\$297 million] was helped by sales of Oculus. The Ocult Quest headset passed \$100 million in content sales in its fir year (Leung, 2020).

Watching classic sport on a laptop or HD video allows you t dip in and out of the content and use other devices – that second or third screening. You can go for a bite to eat and bac without missing too much of the action.

Now try watching a sports match on a VR headset and it's work in progress experience.

Many of the more affordable headsets such as Oculus Go don yet have the picture clarity to match the 2D HD experienc And if you want to do anything else (like use a smartphone you have to take the headset off, creating a disjointed experience.

Whilst people may increasingly be prepared to pay for the VR hardware and gaming content, it is less clear they would invest time and money in VR events such as sport. Sky and Netflix are offering VR experiences and sometimes this is a value added that you get with a subscription rather than paying for a standalone match.

We know that sporting content is also of great interest to Facebook – but the money in VR content currently seems to be with the gaming interactive experiences rather than more passive watching sport.

VR is going to need to become even more immersive to get traction in sports. Manchester City FC and Juventus were two of the first clubs to create Oculus app experiences and Liverpool FC have invested in a VR experience to enhance their live club tours.

Creating behind the goal or pitch side 3D video content is a cheaper (but less satisfying) experience, but creating multi-camera views, apps and virtual fan experiences with physical venues for virtual fan experiences would carry more investment. We may be heading towards the world of another futuristic example from *μSFP*:

"Summer was ready for the final of the 2026 World Cup Fina. She could hear the roar of fans from the tunnel of the Rose Bowl stadium in LA. As she walked down the tunnel with her team mates, she adjusted her headset and wondered what the fans on the pitch would make of her avatar's new pink hair."

One of the closest so far: NCCA, who worked with Intel on a VR app, so you could watch basketball in VR from 5 camera angles for around $8 for six games. They also built in social interactions with other fans, which is also seen as an important aspect of VR in sport, same tech from Intel was also used on the Winter games. Whilst the exact figures were not released it was hailed as a success in the media.

Software Upgrade for Sports

In Summer 2020, sport started resuming in different countrie – the Bundesliga in Germany behind closed doors. Fans wer happy to watch these games on TV or social media in bigge numbers than they would with VR. But with a growin audience and improving tech for VR sporting content, thi could become the new normal.

Immersive video and experiences have potential to retain wha fans like about the sporting experience including secon screening and social interactions (virtual and in person). VR c XR (Extended Reality) will play a bigger part in spo consumption, but for now, headsets are some way off in term of the technology and content. The future may also rest in Al (Augmented Reality) glasses or holograms.

Then beyond headsets and augmented reality glasses, a fina example taste of the future may be found in *Star Trek, Th Next Generation*, set in the year 2364. On board the US Enterprise is 'The Holodeck'. This small XR space can be use to create simulated environments across space and time.

For example, Mr. Data in his spare time becomes Sherloc Holmes, interacting with real and simulated characters an solving mysteries in a simulated London in 1880.

The Holodeck requires no headset, no glasses and is limitle in its space and potential. Furthermore, the danger settings ca be adjusted - so that real characters can become harmed b virtual settings. When this setting malfunctions, storyline abound, and it raises important questions about blending th real and virtual. The Holodeck gives us a glimpse of what th future of XR may look like – a world of infinite possibilities.

About the contributor

Dr. Alex Fenton is a Doctor of Digital Business at The University of Salford Business School. His research is focused on how innovative digital transformation and technologies can improve business processes and impact positively on individuals and sports clubs.

Prior to becoming an academic, he ran a digital development company and helped to get Manchester online with the Virtual Chamber in 1997.

He is winner of a range of teaching and digital industry awards including innovative lecturer the and European Search Awards. He also created a smartphone project for major sports clubs called Fan Fit, which helps fans to get active as an official club app. He also released a book with Routledge in 2020 called Strategic Digital Transformation.

References

Dean, S. (2020) 'The future of football without fans? Danish club Aarhus introduce "virtual stadium" for spectators to attend via Zoom.' *The Telegraph*. [Online] 22nd May.

[Accessed on 26th May 2020] https://www.telegraph.co.uk/football/2020/05/22/future-football-without fans-danish-club-aarhus-introduce-virtual/.

Gerard, R. W., Alex, F., Wasim, A. and Phillip, S. (2020) 'Recognizing events 4.0: the digital maturity of events.' *International Journal of Event and Festival Management*. Emerald Publishing Limited, 11(1) pp. 47–68.

Leung, D. (2020) *Oculus surpasses $100 million in Quest content sales | TechCrunch*. Tech Crunch. [Online]

[Accessed on 26th May 2020] https://techcrunch.com/2020/05/18/oculus-surpasses-100-million-in-quest-content-sales/?guc-counter=1&guce_referrer=aHR0cHM6Ly93d3cuZ29vZ2xlLmNvbS88&guce_re ferrer_sig=AQAAAJMa_ejS4SpnBeZKn1FPvCVChseOT_MBy7ZEUxv6oVoKjb_ nxV3sngml9z_4U4ZpnoPoPcv9zJVOcUE805vvvec.

Chapter 8

Software: Video Tools Overload

Dr Gordon Fletcher

Summary

Not all video conference software was born equal. Differences in functionality - from Blue Jeans to Zoom, via Google meet, Teams and many more - have become obvious, challenging and even embarrassing.

Why software choice matters

Software is not neutral. It contributes to the creation of meaning and shapes the experiences we have through its use. All software acts as a filtering conduit on how and what we do.

Not only are these well recognised observations that can be regularly found in academic literature – such as that of Wiedenbeck & David writing in 1997 in the *Journal of Human-Computing Studies*. They are what a 'user' intuitively knows the moment their IT department invites them to an 'induction' for a new video conferencing tool. The fact that it takes an hour or more to learn how to use the software – rather than being instantly intuitive - reveals everything you need to know about its design consistency, menu interface and exposed complex functionality. Yet despite everyone knowing software is not a neutral agent, companies often make software choices without considering whether the users will enjoy it.

Software choices matter. Large firms usually begin with systematic criteria, like Gartner's, which acknowledge the full range of internal agendas. and the underlying vision, of the vendor. Then they boil the process down to cost, service availability and a feature list. Small businesses and individuals usually buy software less systematically – based on the preferences of close contacts, or the clever marketing efforts of one vendor over another.

Nowhere has the influence of software on our digital interactions become more apparent than with video conferencing. The global pandemic of early 2020 required rapid pivot to home working. As well as competing for space in the living room, organisations needed to select a video conferencing software, at pace. These decisions were completed without time to reflect on the technical or sociological implications. The lucky home workers - in larger and more stable organisations - who were not furloughed had already had the decision made for them by an IT department. For smaller businesses, and for those socialising, there was natural gravitation to the free options as the most important and often sole, criteria for selection.

Accelerating choice in video tools

The choice is accelerating. Wikipedia's comparison of video conferencing software captures the developing feature set. Reaching back through the edits of this one page reveals over 70 edits of all types during the first quarter of 2020. The 2 edits prior to 2020 take the reader back a further three years to 2017. The incremental changes reveal a general trend for all software to coalesce around the same features. The key metrics, including the number of attendees and the size of the grid-view, confirm the same movement to a norm. But general comparisons tend to ignore unique features. For example, BlueJeans uses "Dolby Voice" to offer audio separation of individual participants that simulates a conventional room.

Software glitches

Zoom rapidly became the icon for pandemic working primarily because of its free offer for organisations – compared with many rivals who only offer limited trial periods. The technical benefits of Zoom were, for most, just a pleasing bonus but they did drive personal word-of-mouth recommendations that snowballed the use of the software – an effective social media marketing strategy. But realisation of Zoom's key technical limitation on its free tier soon introduced the world to *conference interruptus*. The forty-minute limit to free Zoom meetings was frustrating for businesses and a disconcerting break in virtual afternoon drinks sessions. Zoom's surge in popularity also tested its abilities. Hackers and security professionals highlighted flaws – reminding IT departments why software selection based on price can be fraught. Passwords were lost, meetings intercepted with pornography and hate messages. Zoombombing became a verb of pandemic lockdown. The risk of video conferencing was highlighted most in the classroom. An indicative – possibly apocryphal - example is of a schoolchild gaining control of a Zoom-based classroom after the teacher's host connection was lost. The child's then repeatedly recited the word "poo" until the teacher could regain control.

And again, software is not neutral. Zoom has made at least some managers change their behaviours. Thirty-minute meetings now figure in the diaries of home-based workers and short, concise meetings are a benefit that will hopefully persist in a Post-COVID world. For the socialite or more committed Zoom user, the solution has brought benefits to the company – doubling its projected 2020 revenues, and funding further service uplifts. Zoombombing is neutralised and the software security improved. The forty-minute limit for users on the free tier remains though. Along with Spotify, it's possibly the all-time best marketing case of a 'freemium' model.

Battle of the grid sizes

The early pandemic that brought success for *Zoom* generated video conferencing war that mirrors the Browser War (Vaughan-Nicols 2018) and the common player has bee Microsoft. The Teams video-conferencing function an integration into Office 365 was for many a little-known aspec of the suite. Now Teams is duking it out with Zoom, and th video conferencing war is about winning over increasingl hardened users with new features and robust capabilities t lock them in.

One feature that may have appeared to be trivial benefit i January 2020 is now pivotal: the size of the attendee grid Teams started the pandemic with 2x2 grid of the last speakers. In April 2020 this expanded to 3x3, then Microso announced 7X7 in June 2020. Not by coincidence, the abilit to see 49 participants matched Zoom's capability precisel April 2020 also saw Teams and Lifesize - another vendor both introduce the ability to raise a hand to attract the meetin facilitator's attention. This feature was in Zoom long befor the pandemic.

The race to introduce more videos streams onto a single scree has meaning and impact. The purpose may be commendable i its ability to create a virtual lecture theatre. The classroom or parliament are some of the few environments in moder society where one individual would be expected to b addressing 49 other people. Few other forms of face-to-fac meetings require this number of participants. Howeve lecturers know there is never a case where 49 people are a equally distanced and framed in front of you. Enthusiast students sit at the front, those trying to avoid eye contract t the back. Having a 7x7 grid of individuals all movin independently without reference to a seated neighbour can als produce sea-sickening nausea.

The danger of a software feature that enables actions not previously possible is that people try to then use it for purposes that it is thoroughly unsuited. A lecture theatre analogy has its purposes in a socially distanced world. The temptation to hold large business meetings because the grid size makes it possible is counter-productive. This scenario may not be within your current experience but the prospect is on the horizon.

The right kind of background blur

Also commonly found across video conferencing software before the pandemic lockdown was the background blur option. In some software, including Teams, changing the background requires placing a suitable background image in a relatively obscure directory not accessible directly through the software itself. With users locked in busy homes, the blur became standard during the lockdown period - a tool that TV personality Trinny Woodall could have used when her partner, Charles Saatchi, walked naked across the camera during a live broadcast direct from her bathroom.

Backgrounds (or green screens on which to digitally project fantasy backgrounds like tropical beaches) provide workers with a critical sense of separation between work and home when such distinctions have become otherwise totally blurred – now an essential tool for any platform. Although the sales of green screens are not recorded as a discrete number in retail sales figures, evidence does exist that the change sales was exponential. A Canadian outlet claimed to have sold the same volume in two weeks as they had in the two previous quarters.

Green screens and backgrounds also highlight a further subtle impact of software's partiality. As new attendees admire the background of another attendee the inevitable question is, "How do you do that?" This is a cue for an impromptu piece of tech support that involves descriptions such as "click on the three dots" or "click on the button in the bottom right hand button." What the helpful volunteer IT assistant often does no[t] realise is that the same software will look different if they ar[e] using the video conferencing software through different we[b] browsers, on a PC, a Mac or through their phone. In man[y] cases there is one priority platform for the software develope[r] while users on the other platforms can only access or even se[e] a more limited set of features. The result is an assumption tha[t] all the attendees on a video conference are in the sam[e] relational *space* and *place* interacting in the same way as ever[y] other participant. This assumption generates specific attitude[s] and approaches to the interaction that would differ if the *priori* assumption was one of heterogeneity. Different variant[s] of the software mediate the interactions in subtle ways. As w[e] continue to use video conferencing these differences continu[e] to compound assumptions in ways that face-to-face meeting[s] do not permit.

So again - software is not neutral. The lucky home worke[r] who are now daily video conference attendees must recognis[e] the mediated differences that software adds to each of the[ir] interactions. These differences shape our view of others and [of] our world continuously.

About the contributor

Gordon Fletcher is Director of Business 4.0 in Salfor[d] Business School. He has recently published with colleague[s] and industry associates *Strategic Digital Transformation: [a] results-driven approach* and the 2nd Edition of *Digital an[d] Social Media Marketing: a results-driven approach* both wit[h] Routledge. Gordon's interests relate to the use of technologie[s] from social and economic perspectives.

Chapter 9

How Choirs Made Perfect Harmony Online

Pegram Harrison

Summary

Choral singing means in-person communication – including of the coronavirus itself, as some choirs have had outbreaks. So what can an online performance of Bach's epic St John Passion teach us about video conferencing in other collaborative jobs – in business and education?

The Choir as a leadership metaphor

I lecture for a living, and sing for fun. My employer is a university business school, and many of my friends are professional and amateur singers. Occasionally, I bring these two parts of my life together by drawing on musical performance practice to teach leadership to MBA students and executives. This sort of experiential learning is a common and effective way for adult learners to enhance skills and gain new insights—for example, about leadership and teams.

But—all of the sudden, all of this stopped.

Lecturing, singing, learning about leadership through singing: none of that could happen in a pandemic. We were truly on mute. When the lockdown first slammed shut, every choir stopped singing. Some were even deemed "super-spreaders," sites of rapid contagion, as the world began to realise how close physical proximity and breathing moistly near each othe in closed-up venues were risky activities.

Some choirs and their extended families suffered terribl tragedies. In one US choir, 45 singers were diagnosed wit Covid-19 and two died. Such acute shocks were hideous, a have all the other deaths caused by this cruel disease. They le to the systematic silencing of an art-form made from sound the forced muting of an activity that is a lifeline of engagemer for amateurs and identity for professionals.

Challenges of singing online

Young emerging singers were hit hard: without a cushion c financial or reputational security, some might never make further up the professional ladder, forced to stop singing at crucial point in their vocal and career development; other would have to earn money in other ways, losing both practice time and contact with agents and audiences. Even fc amateurs, with an activity favoured mainly by older people higher risk for the disease, the end of the lockdown was a uncertain target: uncertainties about rehearsing, or attractin wary audiences wouldn't go away any time soon, and woul only begin to recede long after a track-and-test regime and/or vaccine arrived to help. The muting effect of Covid-19 woul last longer among choral singers than in many other spheres.

But singers and lecturers alike displayed extraordinary copin mechanisms as well as inventive creativity. Though everyon found it almost impossible to sing online because of delays an interruptions and low audio quality, singers persevered wit entrepreneurial work-arounds and artistic creativity to mak new music in new ways. Lecturers did the same. We did n simply record conventional classes on Zoom, boring ou distributed students to death, and instead engaged with the ne possibilities of tele-teaching, some of which are decidedl more powerful than the actual "old-school". The best of u won't fully go back again to what we were doing offline. There were even experiments combining singing and lecturing that emerged from the crisis.

The Oxford Bach Soloists[1] is a fairly new semi-professional ensemble. "New" means "vulnerable" to this crisis: OBS hadn't developed the financial stability or brand resilience to ride out the storm, as some well-established ensembles were better prepared to do. There was no back-catalogue of recordings to keep former fee-paying customers connected. On top of the problems plaguing all performing ensembles, the operating model of OBS made it particularly precarious. It was established to sing the complete vocal works of Bach in order, but a hiatus of unknowable length undermined the ensemble's *raison d'etre*. OBS was established partly to provide performing opportunities for young professionals just launching their careers, but the generous donations funding annual choral scholarships (stipendiary posts, removing the pressure to find the next gig) could not be used during lockdown. Thankfully, donors were patient, redeploying grants to support musicians who would otherwise be out of work, and funding other activities to maintain engagement with audiences: for example, revenue-raising performance opportunities in virtual venues, and making music integrated with technology. One such activity was particularly exciting: performing one of Bach's most challenging and beloved works during lockdown, as if live, in a new, virtual, and entirely exhilarating way, the "St John Passion from Isolation".[2] As a singer, I'm in awe at how this vigorous and impressive performance emerged from the lockdown. As a lecturer, I'm inspired by the leadership lessons it represents, and the learning opportunities it provides.

Technical solution for choirs online

The "St John Passion from Isolation" project was the brainchild of tenor Daniel Norman, his colleagues at

production company Positive Note,[3] and OBS Artistic Directo[r]
Tom Hammond-Davies. Noting that Good Friday 2020 "migh[t]
very well be the first without a sung live Passion since th[e]
Middle Ages," the project began in an effort to raise money fo[r]
Help Musicians UK. It involved a huge and disparate compan[y]
of performers and technicians, including instrumental player[s]
and singers from around the world each of whom wa[s]
physically isolated from all the others. Performers were sent [a]
video of Dan conducting, plus a backing track to listen to i[n]
headphones while playing or singing their own parts into [a]
camera. Dan then aligned and edited the files, stripping th[e]
sound from the video before sending separate strands to Chri[s]
and Jeremy, who adjusted pitch and pace (a herculean effor[t]
and recombined the strands; together, the team made variou[s]
arrangements of video tiles appropriate to the words an[d]
music. There are advanced audio settings in most vide[o]
conferencing systems to adjust the level of background nois[e]
suppression, so that one can listen and sing at the same tim[e]
without provoking as much interference from the AI behin[d]
the scenes. YouTube provides a free uploading platform, an[d]
releasing episodes in stages instead of one solid block mad[e]
the output more dramatic and more feasible to produce, as we[ll]
as opening a longer window of opportunity for fundraising. I[n]
compensating for the loss of live performance, technologie[s]
now within easy reach of anyone, coupled with entrepreneuri[al]
leadership, have made new kinds of performance possible i[n]
ways unimaginable even a few months ago.

Bach's St John Passion narrates the death of Jesus in variety [of]
styles: a solo tenor Evangelist accompanied by small organ an[d]
cello tells the crucifixion story in the words of the Gospel [of]
John; these few tracks were relatively straightforward to recor[d]
and mix because they involved few people. This same sma[ll]
"continuo" group, which is normally the basis of most Baroqu[e]
vocal music, also provided the pre-recorded backing tracks fo[r]
other performers over which to sing their own parts. Choruse[s]

dramatize the story in turbulent, complicated ways; these were difficult to knit together from the isolated contributions of the performers, but also provided an opportunity to add visual effects, emphasizing elements of the music. Contemplative and personal arias are performed by solo singers and *obbligato* instrumentalists—who in real life are accustomed to taking interpretative liberties standing next to each other, and found ways to coordinate emotions and expressions even at distance. Bach's famous chorales—hymn-like pieces in expressive four-part harmony—provide signposts for the abstract theology and personal piety of the piece; these were thrillingly sung by the separate members of separate ensembles for separate chorales, and yet converged with apparent seamlessness through expert audio engineering.

Still—does technical mastery fully explain how all this was done? Indeed not; beyond technique are the passionate motivation of the performers, the delicate and mesmeric gesticulations of the sign-language interpreter for the chorales, the close-up camera angles bringing performers much closer to the audience than otherwise possible, and the inspirational leadership that one senses behind the whole effort, pulling art out of isolation. The videos even stretch far beyond the normal boundaries of musical performance: a visual artist, Paolo Troilo, appears periodically throughout, making a painting in real time, a stark representation of the body of Christ, black with sharp accents of colour which we see being added onto the image, flowing from his pierced side: vitality in mortality.

Superconductors

I'm not sure one can actually teach leadership; like creativity, or poetry, or spirituality, it can be studied and practiced, but being embedded in individuality means that it is not a transferrable skill. But people can *learn* to be better leaders, and how to cope when leadership is "thrust upon them"; so there is a purpose in creating the conditions within which that can happen. That's the value of experiential learning: people draw on some new challenge to re-evaluate what they are capable of, and push themselves further.

Putting a roomful of bankers or corporate executives into the middle of an orchestra, for example, and finding ways for them to interact with entirely unfamiliar processes of production, coordination, and direction is an experience most people don't soon forget, and everyone seems to enjoy. On a smaller scale, and often with more personal impact, bringing together managers and choral singers enables learning about listening, motivating teams, and delegating authority to expertise. Many such experiential learning opportunities exist—using jazz drumming, cooking, climbing, etc. None of them works in lockdown. But the audio-video material of this Bach Passion is a particularly useful resource. It demonstrates up-close the process of organising a complex endeavour, while giving an almost intimate view into the lives and capabilities of the performers. The barriers are down: no one is in concert dress; the conductor is often not visible; the audience is brought right into the performers' homes, even bedrooms. Here are people exercising skills, together; despite being physically separate, they seem close because of their shared activity, an evident passion for music and music-making.

Imagine being able to produce that level of human commitment and engagement in any work-force or team! Just seeing it, hearing it, feeling it is more than inspirational– Bach's epic work seems within reach, feasible. This Passion doesn't explore the Resurrection, but that very absence sends a strong message that human capability can rise above any adversity, even of its own making; we can cope with crisis and compensate with creativity. That is a lesson any leader needs to witness, and a challenge to embrace. No amount of lecturing with PowerPoint slides can possibly covey such lessons as potently as this Bach "Passion from Isolation".

About the contributor

Pegram Harrison teaches at Oxford University, at the Saïd Business School and Brasenose College. After studying literature at Yale, he was a choral scholar at Clare College Cambridge in the early 1990s, and continues to sing in choirs around Oxford. He teaches an MBA class called "Leadership Perspectives from the Humanities", exploring leadership through the performing arts, literature, history, philosophy, classics and architecture.

References

1. www.oxfordbachsoloists.com
2. https://www.oxfordbachsoloists.com/event/st-john-passion-in-isolation/
3. www.positivenote.co.uk

Chapter 10

Business: Making Out of Office work

Angie Moxham

Summary

Business is about relationships – and in moving meeting online for coronavirus, we all became instant students in the non-verbal cues of remote video communication. From a PR professional's perspective what are the practical steps we can take to manage the switch? How can employees manage and represent their lives behind the camera ?

This new normal was a long time coming

Flexible and home working and video conferencing are no new phenomena – but now they are enjoying their X Factor moment becoming the *de facto de rigeur* of commercial and social communications.

I was hired by British telecommunications giant BT in the late 1990s to promote the benefits of mobile working - with brick sized mobile phones and biceps-enhancing laptops. Then, in the early noughties, Skype hired us to promote their "revolutionary", (it was, to be fair, though they have drifted since), voice and video over IP conferencing platform. So Zoom and Microsoft Teams as concepts are not new; just on better tech steroids. Yet it has taken a pandemic for companies and their employees to truly embrace the "new normal" out of necessity – and now there will be no going back.

Corporate coping strategies

So how have we coped as individuals with these changes ? How have companies evolved ? Will Out of Office working outlast the coronavirus ?

In a recent survey by Deloitte, the majority of City workers expect to spend more time working from home after the lockdown has ended. The study concluded that more than three quarters of financial services employees believe that they will work remotely at least one day a week after restrictions are lifted; 43 per cent expected to work from home for more than two days a week; and only 10 per cent of respondents said they'd found operating remotely a negative with 70 per cent saying it had been a positive experience.

That's a transformative vision of how the workplace will function for employees. But how do employers feel? There are clearly financial benefits to be had for some, savings on office overheads and business travel expenses, for example. However, historically, many leadership teams struggle with trusting their employees to be fully productive when working from home; and colleagues who are required to be in the workplace often resent flexible working practices if they're not universally offered.

In my experience, the younger and entrepreneurial businesses have always embraced operating "out of office" far more than larger, more traditional corporates. Partly because the latter's workforce are typically digital natives as opposed to digital immigrants. And smaller ventures have less of a need or desire to control their workforce. in a static office. My businesses, like most, arguably, are relationship businesses.

It may, however, not matter what business leaders and owners feel about post-lockdown working practices; change may be forced upon them. People living in Britain could reportedly soon be given the legal right to work from home, as the government's long-term plans begin to take shape. Whitehall

officials at the Business, Energy and Industrial Strateg.
department are said to be considering introducing legislation t·
allow staff to operate remotely after restrictions are lifted. Th·
decision would make it easier for people to socially distance
and reduce costs for employers who will be expected to mak·
office spaces safe. Employers would only be able to reject
working-from-home request if a staff member's job can onl·
be done in the workplace, according to reports in *Th·
Telegraph* in May 2020. Ministers allegedly supporting th·
plan are hoping it will limit congestion on public transpor·
One said: 'It makes complete sense.' Trades Union Congres·
general secretary Frances O'Grady also claimed the right t·
work from home would be a 'big step forward' for workers.

Whatever the outcome, businesses and PR consultancies - lik·
mine - should look to learn from the good, the bad and th·
downright ugly of what home working via video conferencin·
has taught us.

So here are my top findings of the benefits and drawbacks of
shift to Out of Office:

The Good

1. **Optimise the Out of Office stage directions**

   Video conference calling from your home creates a much
   more personal and intimate ambience. How much have w·
   enjoyed the non-verbal, only partially-curated "through th·
   keyhole" insights into our colleagues' homes? We gain
   subliminal insight into many potentially helpful
   dimensions of their lives, from their choice of décor,
   selection of books, home clothes (from the waist up); all
   transmit interesting data about their attitude. Dogs barking
   kids shrieking - the Zoom experience personalises busines·
   meetings, reduces formality and arguably makes them not
   only a more enjoyable experience, but a better exchange c·

values than any meeting in the deliberately anodyne conference room.

2. But Stop Watching

People inevitably feel more self-conscious on video conference calls – because they are self-aware. I've found it hugely disconcerting having to see my face amidst the gazing gallery. Where do I look? How do I look? OMG look at my hair?! Because of that, meetings are far more efficient. For the most part, people just want to get through the agenda and be able to switch the camera off. The 40 minute limit of free Zoom account, (the staple of start-ups and home use), has taught us all to be more focused and effective in our conversations.

3. Shoot the commute

According to a Lloyd's Bank study, Britons spend 492 days of their life travelling to work, at a cost of £37,399. These hours are substantially lost, the money wasted, the environmental damage done, the health impact banked. Whilst most of us try to use our commute productively, working on a train is far less efficient than being able to glide seamless(ish)ly from shower to study. Aggregating the collective efficiencies of more time spent out of office, reducing the 70 weeks' commuting time, is a surely a national economic and psychological win win?

4. She said, he said

Where there are people, there are always politics. My biggest business so far was PR Agency 3 Monkeys, which I sold in 2016, with a team around 85 strong. When asked what I did for a living I always replied: "I run a kindergarten, ages 23-60". One of the biggest headaches of running any company is dealing with the people and politics which inevitably and exhaustingly feature day in and day out of office life. Remote working cuts through the insecurity. An absence of gossipy "water cooler" moments, or whinging over stale pints at The George after work can drive productivity. And people appreciate one another more due to their absence.

The Bad & The Ugly

5. **Video killed the radio star**

Communications via video conferencing challenge the listening culture, which is fundamental to any enterprise's success. When we're all one Microsoft Team, it's hard to do more than say your piece. Attention to body and facial language is far harder in the two-dimensional landscape of a video conference. For the introverted colleague ,(often the most valuable kind), getting an important point across amidst the jazz hands of the virtual meeting room is a challenge. So make a note of who's not saying much and call them after the meeting to listen to what they might have wanted to say, and them incorporate that into any feedback notes you collate.

6. **Where's the off switch?**

Whilst some may find structuring their home working day easy, for single Mums like me, it's hard. With three grumpy teens to motivate and supervise in their virtual education, let alone shop, feed, wash and clean our home, it's hard to get into a rhythm of work that doesn't mean you're regularly burning the midnight oil. As an entrepreneur and someone who had to write proposals with one hand whilst breast feeding all three through the night, this doesn't phase me. But for many, this has become a real issue, adding to the anxiety many, understandably already feel.

7. **It's good to talk**

When we're all in a shared place of work, we're in constant convo mode, sharing ideas, issues and enjoying general banter. When locked down, we're communicating more online which has always been a personal bug bear, an area which businesses need to seriously review and improve. Email was invented to allow us to send large documents through the internet rather than standing by a fax machine for hours on end. Yet it's become the de facto business communications method creating unnecessary, inefficient, day extending, arse covering, non-decision making, political paddling reams of diatribe. Email schmemail. It kills companies, cultures and is, for the most part, joyless. If there's a conversation to be had, have it. On the 'phone or via video conferencing.

We will wait to see if the period of global Out of Office gives birth to a new way of working for businesses. It's hard to predict what the long tail of this will be, aside from any possible legislation which could force companies into a "new normal". As a business owner, I've always encouraged this new age working; why wouldn't I? The health of my team is of paramount importance; we only live once and I don't believe anyone can enjoy a truly fulfilled life/work balance in a clock on and clock off environment, whether you have children or not.

For what it's worth my view –and I don't mean it cynically – is that human beings will just revert to type at home and at work. We can't help it. We're wired that way.

But my shout out to all businesses, irrespective of your size or sector, is that we all seize the opportunity to do a proper root and branch review of all of our working practices and business comms. Consult our workforces and find out what's the best of all our worlds to maximise health, wealth and happiness for your organisation and home lives since one feeds the other. And focus on real and necessary communication, face to face,

virtual or otherwise. All media and channels have their rightfu
place. It just takes trust, common sense and a lateral and libera
perspective. If you do, you could reap huge rewards in you
reimagined business and work lives. Good luck and share you
thoughts and experiences.

About the contributor

Angie Moxham is one of the best known female PR gurus i
the UK industry, having created three highly successfu
communications agencies. Since 2018 she has been founde
and CEO of The Fourth Angel. She previously run two awarc
winning agencies – Le Fevre Communications and 3 Monkey:
which she sold to the Edelman Group in 2016. During he
career she has worked with everyone from Microsoft, Coc
Cola, WRAP and Sainsbury's, to United Biscuits, Superdruş
Starbucks and Disney to name a few. She has an Englis
degree from St John's College, Oxford.

BITE-SIZED
BOOKS

Bite-Sized Lifestyle Books are designed to provide insights and ideas about our lives and the pressures on all of us and what we can do to change our environment and ourselves.

They are deliberately short, easy to read, books helping readers to gain a different perspective. They are firmly based on personal experience and where possible successful actions.

Bite-Sized Books don't cover every eventuality, but they are written from the heart by successful people who are happy to share their experience with you and give you the benefit of their success.

We have avoided jargon – or explained it where we have used it as a shorthand – and made few assumptions about the reader, except that they are literate and numerate, and that they can adapt and use what we suggest to suit their own, individual purposes.

BITE-SIZED BOOKS

Bite-Sized Books Catalogue

We publish Business Books, Life-Style Books, Public Affairs Books, including our Brexit Books, Fiction – both short form and long form – and Children's Fiction.

To see our full range of books, please go to https://bite-sizedbooks.com/.

Printed in Great Britain
by Amazon

43585970R00050